ADOBE UP CLOSE

The Beauty of Buildings Made of Mud

ADOBE UP CLOSE

The Beauty of Buildings Made of Mud

MARCIA JOHNSON

PHOTOGRAPHS BY MICHAEL GAMER

SUNSTONE
PRESS

SANTA FE

Sunstone books may be purchased for educational, business, or sales promotional use.
For information please write: Special Markets Department, Sunstone Press,
P.O. Box 2321, Santa Fe, New Mexico 87504-2321.

Design › R. Ahl
Printed on acid-free paper
∞

—————————————

Library of Congress Cataloging-in-Publication Data

Names: Johnson, Marcia, 1941- author.
Title: Adobe up close : the beauty of buildings made of mud / by Marcia
    Johnson ; photographs by Michael Gamer.
Description: Santa Fe, NM : Sunstone Press, 2021. | Summary: "A review of
    adobe as used in regional architecture, especially New Mexico, where
    adobe is an ancient and dynamic use of mud and dried mud sections that
    create a house that is sensual and charming"-- Provided by publisher.
Identifiers: LCCN 2021035668 | ISBN 9781632933539 (paperback) | ISBN
    1632933535 (paperback)
Subjects: LCSH: Building, Adobe--New Mexico. | Building, Adobe--Colorado.
    |
    Vernacular architecture--New Mexico--Santa Fe. | Vernacular
    architecture--New Mexico--Taos. | Vernacular architecture--Colorado.
Classification: LCC NA4145.A35 J64 2021 | DDC 666/.737--dc23
LC record available at https://lccn.loc.gov/2021035668

—————————————

WWW.SUNSTONEPRESS.COM
SUNSTONE PRESS / POST OFFICE BOX 2321 / SANTA FE, NM 87504-2321 /USA
(505) 988-4418 / FAX (505) 988-1025

"...building in adobe is like a disease; once you start using it you can't really ever stop."

—Georgia O'Keeffe, quoted by Franklin Israel in *Architectural Digest*.

"...under the ineffable alchemy of the sky, mud turns ethereal and the desert is a revelation."

—Charles Lummis, *The Land of Poco Tiempo*

CONTENTS

HOW *ADOBE UP CLOSE* BEGAN

Back in the day women had two primary choices for a career: teaching and nursing. Although married women didn't usually work it was a good idea to have a backup plan. I chose the road probably most travelled: teaching. In order to keep a teaching certificate current one needed continuing education credits. I learned that Adams State College, (now University) offered the requisite credits through a course in water color painting in Taos, New Mexico—a one week intensive.

It was affordable, more interesting than time in class and made possible for me by a team of grandparents and the willingness of husband/father to take over my parental responsibilities while I was gone. While in Taos the class went into town or to spots out of town that were primarily of adobe homes sheltered by cottonwoods and set under the bluest sky. I came home as unaccomplished a watercolorist as ever, but infatuated by the area.

I looked for a book that I could pull from the shelf whenever I wanted to revisit that wonderful week. There was no book. But there should have been. About this time, I met Michael Gamer, who had the skill, the time, the interest and the equipment to take the pictures for the book that should be and that is when most of the pictures in this book were taken.

Just as I was approaching publishers *Santa Fe Style*, by Christine Mather and Sharon Wood came out. This beautiful style book of 250 pages in full color took the reader in a different direction, but addressed the niche I was hoping to fill.

I went on to work as an interior designer and Michael moved as well. After over twenty years in design, I ran for public office in Denver and served 21 years in three different positions. The would-be book stayed shelved.

And now, after these many years, the book that should be is here: *Adobe Up Close*. Books about the charm and challenges of adobe have come and gone out of print. They live on my shelves along with almost every book written by Frank Waters, who captures the spirit of the area as none other.

INTRODUCTION

The Spanish Conquistadors, coming north from Mexico, found pueblos in what we now call New Mexico that seemed familiar, reflecting the Moorish architecture the Conquistadors knew well. *Adobe* is the Moorish word for *"at-tub"* or *"al-tub,"* meaning sun-dried brick.

At the Taos Pueblo the Conquistadors found a community living in four-story apartment houses that had been built long before settlers came to establish colonies in the East. Between two large buildings was the plaza where outdoor space was, and still is, used for cooking, caring for children, and ceremonial dancing. The French also were vying for settlement, each in the name of their respective countries. Frenchman Bishop Lamy, in *Lamy in Santa Fe* by Paul Horgan, speaks of finding the houses made of earth that merged into the ground.

The next colonizers were the artists who found in the village of Taos the quality of the light and the Indians as inspiring subject matter. The original seven Taos Society artists were Henry Sharp, Irving Couse, Robert Henri, John Sloan, John Marin, Victor Higgins and Marsden Hartley, soon followed by many more. Henry Sharp says, "I was the first to carry a paint box into Taos Valley." Their work is found in museums and quality private collections. The artists were preservationists who lobbied the legislature in Santa Fe to adopt the territorial style of adobe construction for public buildings. Coming decades later, in the 1900s, people found the building material adaptable to the styles they were familiar with.

Why does this book focus on southern Colorado and northern New Mexico when adobe has been used throughout the Southwest and there are good examples all the way to San Diego? Because the most important prototype, the Taos Pueblo, almost a millennium old, is there; because the massive iconic San Francisco de Asís Church is there; because the use of adobe can be seen in its simplest, humblest form; because there is a course of study on adobe construction offered at the Vista Grande Charter High School; because the area is a magnet for scholars and tourists; and because the making of adobe brick is being commercialized due to the demand for these bricks. Some of these factors apply elsewhere, but not all in one succinct area.

IN APPRECIATION

Only through the hospitality of the people who invited us into their homes was this book possible. With tripod set up we would be taking an outdoor shot when the owner would invite us inside for more pictures. We might rearrange furniture and make changes to get the best light. The equipment was bulky and sometimes we were working in close quarters surrounded by family treasures. After using up a couple of hours our host would say, "You should see our friends' homes." They would make a couple of phone calls and send us on our way.

Special thanks to the late Mrs. Hugh Catherwood and to Marilyn and Charlie Batts, who provided us with places to stay. Both homes are represented in the book. High school classmate Judith Houser Gritz told of the remaining walls of the storage rooms at her house.

My husband, Will, made innumerable trips from Denver to New Mexico. He enjoyed the scenery, the places we stayed and the food. More family help: my grandson, Mark Moreno, gave me computer support, making this manuscript transmissible.

Thanks to Bobbi Barrow, Tom Carr, Carol Fertig, Martha and James Hartman, Michelle Salisbury and Cathy Wright, who served as readers and content advisors. Ms. Wright has worked most recently at the Albuquerque Museum of Art and History and the Colorado Springs Fine Arts Center. She hung many exhibits of related topics in museums in Colorado and New Mexico and published extensively on related topics. Matthew Restall, Edwin Erie Sparks Professor of History and Director of Latin American Studies at Pennsylvania State University, made helpful suggestions. Errors that remain are mine.

See more names of Those Who Helped following Suggested Readings.

ADOBE SEEN WALKING AROUND SANTA FE, TAOS, NORTHERN NEW MEXICO AND SOUTHERN COLORADO

Where can one find these buildings? Almost all of the places in *Adobe Up Close* were photographed in Taos and Santa Fe, New Mexico and in nearby villages and in southern Colorado.

"New Mexico, like the dearest women, cannot be adequately photographed...one cannot focus upon sunlight and silence; and without that the adobe is a clod. ...Under the ineffable alchemy of the sky mud turns ethereal." Charles Lummis in *The Land of Poco Tiempo*.

One can wander the streets of these towns and see some of what is shown in these pictures, but the photographs were taken between 1985 and 1988. People in adobe homes have an itch to keep changing what they have. It is so easy to do. Thus, neither the author nor the photographer is sure we could find our way back.

One is probably only able to see what is shown here exactly if a property had been protected by historic designation at the time it was photographed for this book.

Most owners asked to remain unnamed. Photographs taken in towns and villages are intermixed.

Adobe brick must have its layer of mud re-plastered every few years. Here the wall shows both well protected areas and vulnerable ones, while the house down the lane is in excellent repair. The little structure above the gate is a bell tower whose resident is missing.

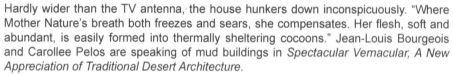

Hardly wider than the TV antenna, the house hunkers down inconspicuously. "Where Mother Nature's breath both freezes and sears, she compensates. Her flesh, soft and abundant, is easily formed into thermally sheltering cocoons." Jean-Louis Bourgeois and Carollee Pelos are speaking of mud buildings in *Spectacular Vernacular, A New Appreciation of Traditional Desert Architecture.*

One can't tell by looking how old a building is. With attention, a wall can last a very long time, but left alone it could melt away in just a few years, due to rain and snow, if it doesn't receive a protective coat. Fortunately for the planet, the abandoned adobe returns to the earth as a part of a perfect recycling process.

Right in town in Alamosa, Colorado a woman built a double-walled adobe storage building, of which only this partial wall remains. In the building the owner stored the fruit and vegetables she canned.

Issue #22 of the publication for Colorado Preservation, Inc., 2019, "Colorado's Most Endangered Places," featured the adobe potato cellars of the San Luis Valley. These are double walled storage sheds, constructed in the 1900s, that held potatoes until the right time to send them to market. One of the major crops of the Valley is potatoes, now supplying 90% of Colorado's production. The profit for the crop would be lost if all of the potatoes hit the market at the same time. The adobe sheds kept the potatoes at exactly the right temperature and humidity until the farmer wanted to sell. Not all sheds had double walls, but that feature added insulation. The concern over the cost of time and labor today in building with adobe means potato sheds are, and have been for decades, built with commercial products.

This building being restored with funds from History Colorado's state historical fund, is a former newspaper office in San Luis, Colorado, Colorado's oldest town.

The horizontal band above the doors is the bond beam supporting the *vigas* of the inside ceiling. The sloped tin roof and stack of piñon firewood help deal with the winter cold and snow. At the high altitude of southern Colorado and northern New Mexico the winters are long and cold. Many people rely on firewood for heat, so it is gathered all summer.

The Martinez Hacienda was a self-contained compound with 21 rooms and two small plazas. One portion was for the family, the other for workers' quarters. When the Indians were on hostile forays onto Spanish lands the people rounded up the livestock and brought it behind the walls.

Located two miles west of Taos is the Martinez Hacienda, circa 1804, the last hacienda remaining in the area. The Kit Carson Foundation is the current owner and it is on the National Register of Historic Places.

The wall is the first layer of protection; the coyote fence is next. The coyote fence was devised to make the enclosure impenetrable, thus protecting the livestock within. The materials for the fence are cheap, but the labor intensive, as this requires gathering, placing and wiring together the sticks that are brought down from the nearby hills.

This coyote fence has thicker uniform peeled sticks. Its texture contrasts with the wall. The texture contrasts with the smooth walls of the home.

This close to the street the owners are opting more for privacy than light.

"It doesn't matter if some of us go one way and some another," say the spirals of these columns. "This is the land where individuals have been idiosyncratic for a very long time," wrote Charles Lummis. When the columns were trees, they may have sheltered Spanish padres 350 years ago.

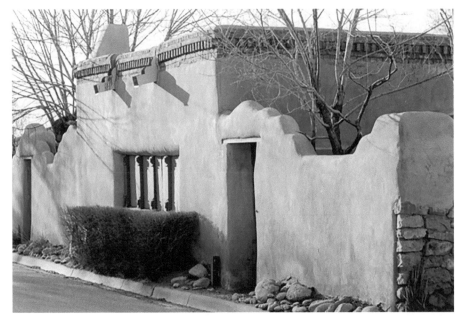

Brick firewall at ceiling height and *canales* (down spouts), deep-set shuttered window, deeper doorway, the exposed lintel above the door and even the rock detail at the base give this fairly new one-story building all the attributes of quintessential adobe style.

Parking on the narrow streets of Santa Fe is a challenge, as is walking out of the traffic, since there are few sidewalks. However, only by walking can you inventory what you pass by: brick fire wall, details of the metal downspout, the soft edges of the wall...so it all works out.

More to notice while walking is the row of skulls on the wall in the background. The cottonwood tree was whacked off and is sprouting again. Perhaps the branches were a threat to the hard-to-repair wall should a load of snow cause a limb to break.

Sit a spell and visit with the woman baking bread and the others who may be going to the plaza to trade and gossip with friends. Artist Fern Ray painted these windows for the El Nido restaurant in Tesuque.

"I like to make tea for my husband and me. At midday we take our tea outside and sit on our bench, our backs against the wall of the house. Neither of us wants pillows... The bench feels strong to us, not uncomfortable. The tea warms us inside, the sun on the outside. I joke with my husband; I say we are part of the house: the adobe gets baked and so do we." From *The Old Ones of New Mexico,* Robert Coles, University of New Mexico Press.

HISTORY OF NEW MEXICAN ADOBE

The Southwest became a part of the United States from a different compass point than the western expansion from the eastern part of the country. This migration came north from present day Mexico and with it the Spanish Conquistadors brought the Catholic faith and a different way of building. Changes in adobe buildings follow the history of the four flags that have flown over Santa Fe. *Taos Adobe, Spanish Colonial and Territorial Architecture of the Taos Valley* by Simms, Bunting and Booth, divides the history and use of adobe into four periods: Indian, Spanish Colonial, Territorial and, later, American.

Adobe buildings from the Pueblo Indians to today reflect this history of architecture. The Indian period is expressed in the pueblos that were built using the dirt at hand, plus water, to make mud balls called wattle and daub. The Spanish conquistadors and friars found 66 pueblos in 1598 in what is now New Mexico. They brought the brick forms of adobe construction they had learned when Spain was occupied by the Moors. There are now nineteen pueblos.

Spanish Colonial buildings are characterized by being very linear and simple. Rooms were added to the ends of previously built ones as needed by growing families. Land was available, so one-story buildings made sense. They were easier to build, helped hold in the heat or keep rooms cool and could enclose a common area, a plaza. The contiguous roofs were a flat playground for the children. A free-standing church dominates every Pueblo.

The US Army arrived in 1846 followed soon after by the railroads. The Treaty of Guadalupe Hidalgo, signed in 1848, transferred what is now New Mexico to the US from Mexico. When that happened New Mexico became a US territory. In came the styles from the east and materials that gave expression to those eastern fashions such as wood filigree, turned posts, sheet metal for roofs, pickets for fences and brick used first for chimneys and then for whole houses. Victorian and Greek revival details seen in pediments over doors and windows were now fancying up adobe homes. The railroads brought paint, wallpaper and linoleum for interior embellishment.

New Mexico became a state in 1912. Beyond adding plumbing and electricity, American-influenced adobe buildings might now have solar panels and use adobe mud augmented with asphalt emulsion for weather proofing, but there were few dramatic architectural style changes...until Frank Lloyd Wright's plans were resurrected.

THE TAOS PUEBLO AND BUILDINGS

INFLUENCED BY THE PUEBLO

The Taos Pueblo is the quintessential Indian building. "Nowhere else in the United States can a style of architecture be found that traces its descent in an unbroken line from aboriginal American sources," wrote Claire Morrill in *A Taos Mosaic*.

The Taos Pueblo, over 900 years old, is a compound of two-four story apartment houses facing each other across a plaza divided by a stream. It is, according to Charles Lummis, in *The Land of Poco Tiempo*, proof of the superiority of the Pueblo Indian who is, "the only aborigine on earth who inhabits many-storied buildings and the only man who ever achieved, in our land, such architecture of unburnt clay." It is built of daubed mud. Small wonder that an entire region pays tribute, through plagiarism, to this way of building.

The Taos Pueblo people share religion and social and economic life. Many of the people living here work outside of the Pueblo, coming home in the evening when they might spend time on traditional crafts. Taos Pueblo pottery is very collectable. Four gentlemen talking with us said that "Men have done the plastering since the time they weren't needed as warriors." In other pueblos plastering is usually women's work.

Other people live in homes on Pueblo land outside the walls and others live far away, but try to get home for important occasions, such as the dances for religious feast days. Originally most men were farmers, but today both men and women might work in fields and both work in every other profession and location.

In the Taos Pueblo, as in the rest of the nineteen remaining Pueblos, is a church representing the faith brought by the friars who came with the Spanish Conquistadors. By the sword conversions were made and the Christian faith remains, in combination with the ceremonies and traditions that honor the beliefs that the Spanish found when they arrived. Here monotheism and polytheism co-exist. On Catholic feast days people will go to church and then on Indian feast days people will dance or watch dancers whose choreography and costumes might be a part of prayers for rain or good crops. On most feast days tourists are welcome to watch. Anglos might ask questions and the people, who are exceedingly polite, will answer with something, but know that what is secret is kept secret.

Oliver La Farge, in *The Door in the Wall*, describes a generic pueblo thus: "...plaza is an expanse of hard trampled clayey earth, dusty in dry weather, with nothing growing but one splendid cottonwood near the middle. The flat-roofed homes—some one story, some two stories high, with some porches—form a nearly solid enclosure. The native adobe is warm brown with a slight sparkle to it, under porches—in New Mexico they are called by the Spanish name, *"portales"*—walls are usually whitewashed. There is a variation in the size of the houses, in the arrangement of doors, and in the windows, which range from tiny pre-historic apertures to (those with) modern metal casements. The whole, though uniformly composed of rectangles, is not monotonous. Adobe weathers into softness and the Indians build largely by eye, so the angles are not quite true and the most regular construction achieves naturally the irregularity that the Greeks used to plan for... One of the much-touted charms is how it takes sunlight and shadow, the sunlight absorbed and softened, the shadow luminous...the Pueblo is snug. You could almost say contained, almost fortified."

The Taos Pueblo located just north of the town of Taos, New Mexico.

Buildings borrow traits from the Taos Pueblo. The Pueblo style gets more sophisticated.

Building codes require the use of the Pueblo style in Santa Fe, which is why the capital city of New Mexico is called "The City Different." Here the walls curve modestly back into the windows, which are relatively small and the base of the building is thicker than the walls above. Approaching daylight plays with the uneven surfaces.

San Francisco de Asís Mission Church in Ranchos de Taos. People have been praying at San Francisco de Asís since 1750. The walls are many bricks thick. Travelers have seen the flying buttresses of Notre Dame in Paris. Buttresses in adobe don't fly. Here they shoulder up to the task of supporting the massive walls to which they are assigned. The buttresses mold the church to the earth.

Taos has attracted artists for decades and each seems drawn to this church to paint, to capture the solidity, to catch light that shows the modulation of the walls. Georgia O'Keeffe was only one of many. You will see people with easels set up, working to catch the effect of light on the walls. In the shops surrounding the plaza of the church you can buy originals and prints of the church.

The walls of St. Francis de Assis are many bricks thick and covered with mud. Winter is hard on the adobe so it must be renewed every few years by people who live in the village of Ranchos de Taos and others who come home from far away to work voluntarily as a show of faith and of community. The adobe flakes and cracks during the winter. The layers discernable here show the addition of adobe year after year. This was traditionally women's work. These days calls go out for both men and women to volunteer.

At one point the church council thought to remedy this situation by coating the church with a hard plaster not made of mud. This was controversial but the priest thought the adobe was a sign of poverty and even a lack of respect to the faith. Plaster was a status symbol. But water got under the plaster and melted the adobe brick, which was discovered when people saw mud running from under the plaster at the base of the walls. The plaster had to be chipped away and adobe re-applied. It is harder every year to find the labor to re-apply adobe.

The support for this gate that looks rather like a buttress.

Materials at hand were piled up to buttress the wall. There may have been a weak spot, a crack or a bulge, although just the heft of the walls justifies the support. "Buttress" is both a noun and a verb.

Early and late light gives emphasis to the modulation of the wall, supported by buttresses like the more massive ones supporting the walls of the Church of San Francisco de Asís. Tucked almost out of sight is a door to the property sheltered from the traffic on busy Canyon Road in Santa Fe. Small cups in the roadway collect a little water for the xeric plants.

The massive buttresses, strong beam and hefty shed doors are made to last for decades, as they may already have. What is stored inside is quite secure.

The door surround looks Aztec. By 1846 the legend that the people from the pueblos were related to Montezuma (or Moctezumas) was widespread. Montezuma became ruler in Mexico in 1440 and it was he who built the aquaduct from Chapultapec to Tenochtitlan. Migrations were prevalent before the European conquest. These travelers were the guides for Coronado.

COLONIAL STYLE ADOBE

TERRITORIAL STYLE ADOBE

Here is an illustration of the transition from Indian to Colonial architecture. The building on the left, with its small dirt courtyard, looks like it once was at the end of a street. The Sena Plaza Gallery looks like it is suporting its elder. The brick cornice on the right, a colonial trait, is an odd fit with the wooden *vigas* from the ceiling inside.

Wood coping is supported by a beam and corbels protecting the top edge from water damage and is a great example of the territorial style. The post supports two beams and a corbel.

People coming by wagon train from the East brought what they knew from home as reflected by this territorial style house translated into adobe. People settling here wanted some representation of civility back home.

The territorially detailed *portale* in El Rito, New Mexico is old enough to have lived through the build-up of years of dirt left from past washouts.

CONTEMPORARY CHANGES IN ADOBE STYLE

Evolution is inevitable. Change is driven by the artistic talents of architects and the predilection of owners who want something a bit more individual or more up to date. The American style would include free form buildings. Melding the requirement for following along, but managing to move along, is a home that looks very 1950s.

A disciplined departure from traditional houses is the Pottery House designed by Frank Lloyd Wright in the 1940s, but not built until the 1980s.

In the 1970s the United States experienced an energy crisis that brought out the missionary zeal of the new converts to this way of building. Not only would building with asphalt emulsion re-enforced adobe be a savings in non-renewable resources, but it would be cheaper, addressing the high cost of housing. It would astonish these advocates to know that by 2020 very little of this transformational construction has spread beyond the region.

A community called Earthship (one word as used by the residents and scholars) near Taos, New Mexico, holds classes and workshops and has interns and apprentices. The information they have on adobe asserts that adobes built partly in the ground and powered by solar is to be found in every state of the Union and in 39 foreign countries. However, not much of North America has yet seen the vision and made the transition,

To see unusual buildings in adobe from the 30% of the world that has always built with adobe, try to find the publication called *Down to Earth* based on an exhibit at the Centre Georges Pompidou in 1980-81. The exhibit went on to museums in Europe, North America and the Third World. Many extraordinary buildings are in adobe in Africa where adobe has been used for centuries.

Arte Moderne of the 1950s is expressed in this home constructed of stucco that is supposed to look like adobe. Glass brick has been used instead of a picture window. The strong summer sun at this altitude brings a lot of heat through a south facing window.

The Pottery House in Santa Fe suggests that architect Frank Lloyd Wright was caught up in the rhythm of the local vernacular. Mr. Wright gave the house the look and feel of a giant Indian pot or bowl because of its earthen construction and its molded rounded contours. His 1941 plans were rediscovered in the mid-1980s and executed by Charles Klotsche.

North of Santa Fe is a sprawling community of solar homes. The goal is to have no heating or air conditioning bills. The homes are spaced to allow for a few goats to provide milk and a henhouse.

Architect John Gaw Meem explained, "Some forms are so honest, so completely logical and native to the environment that one finds—to his delight and surprise—that modern problems can be solved by the use of forms based on tradition."

Contemporary adobe home built into a hillside was a way to maximize the benefits of the earth holding in heat in the winter and shielding the people from the heat of the sun in the summer. The windows face south to gather in the winter sunshine. Scattered throughout are hot tubs.

The owners of homes in this contemporary development near Santa Fe aspire to heat without technology. South facing windows gather the warmth of the sun which is stored in a solid wall in the room just beyond the windows, called a trombe wall, which will hold the heat.

Earthship Biotecture Visitor Center.

The slanted windows are protected from the sun by window coverings that can be lowered if there is too much of a good thing. Another way to stay cool is represented by the room at the far left that is recessed under a *portal* (porch) roof and stays in the shade in the afternoon. People focused on collecting passive solar heat in the 1960s and 1970s when they began building with self-sufficiency in mind. Tax incentives have made the use of solar panels more prevalent.

This earth home, made of adobe and bottles and cans collected as a means of recycling, captures the sun for heat and energy generation. The people living here rotate on a casual basis in commune fashion. Not seen here, but in other buildings in the compound, used tires are a part of the construction, both to recycle the tires and to capture the heat released by the rubber. Banks of potted plants inside and gardens outside provide the food the vegetarians are committed to.

As stated earlier, people are always changing their adobe homes and here an update probably makes this home totally self-sufficient. The home is not too large, so these panels may be enough to gather in the sun's energy to be all that is needed in the sunny but cold winter. A stylized sun is New Mexico's state symbol.

--

THE PROCESS: MUD IS NOT A DIRTY WORD

People would build their own homes from bricks made on site. Students at the College of Northern New Mexico learn how to make bricks that meet the standards for certification by the Earthbuilders Guild. They can then build their own homes and commercially help meet the demand in the area.

Adobes up close, stacked and ready to be put to work. They are made in forms resulting in bricks, usually 17" x 10" x 7". Originally building with adobe meant freedom from the cash economy. Having pre-made bricks delivered to the site is the norm now and not inexpensive.

Adobe buildings come from the ground itself. The digging begins on site, massing enough dirt to make the bricks, which is done by mixing dirt with water and adding some straw, then mushing this into forms for making usually four bricks at a time. The straw helps hold the mud together. Not all clay makes good bricks. The bare bones of this plaster wall, an addition to the wall at the far right show the adobe before and after a coat of plaster has been applied to the brick.

"Adobe," a word of Arab origin, means "earth from which unburnt bricks are made," explains Bainbridge Bunting in *Taos Adobes: Spanish Colonial and Territorial Architecture of the Taos Valley.*

Here the commercially made bricks were being used in building the home of former Secretary of the Interior Kenneth Salazar and his wife, Esperanza (Hope) Hernandez Salazar, in the San Luis Valley in Colorado. Secretary Salazar also served as Senator from Colorado. What might it have been like to be living in Washington DC, in a city of marble and stone, while planning their next home to be made of mud? The Salazar family has had land in the San Luis Valley for six generations. Secretary Salazar later became Ambassador to Mexico.

Spain, then Mexico, granted land to those who would cultivate the land in Texas, Arizona, California, New Mexico and southern Colorado. The villages in the mountains and valleys of northern New Mexico and southern Colorado were so remote that the language there was spoken with a Castilian accent into the 20th century. Interestingly, there are some Jewish traditions that have survived among the largely Catholic population.

The adobe texture on the cover is from a brick on this site.

The windows of the Salazar home being framed in. Clouds are hiding the view of the Sangre de Cristo Mountains on the east side of the San Luis Valley.

Supplying adobe brick is an ongoing enterprise. To supply the bricks necessary for new construction commercial brick yards make the bricks in three categories: all mud, mud augmented with partially emulsified asphalt, and brick augmented with fully emulsified asphalt. The asphalt adds weatherproofing qualities. After drying they are ready for use.

Fully stabilized brick, referred to by the New Mexico Building Code, contains enough stabilizer to limit water absorption to 2.5%. To achieve this, bricks are manufactured with 5% to 12% asphalt emulsion and are completely waterproof. Semi-stabilized brick is water resistant and can also be made and stockpiled because it is rain and snow resistant.

After layers of adobe brick are laid to form the walls, the exterior mud coat is made of a thinner mud called slip, which is thrown or slapped onto the wall. After the mud is troweled onto the wall or thrown in mud balls, it is smoothed and may be rubbed with sheep's skin, adding luster from the lanolin in the wool. Great to have a team member who can handle the lower areas and save the adults from crouching uncomfortably.

Although the wall has its coat on, the hem is too short. Splash-back has affected the base. The pebble, which should have been felt and discarded by the person doing the plastering, has created a vulnerable spot.

Straw helps hold the mixture of dirt and water together. Speculation is that because the straw gleams in the sunlight it was this that excited the Spanish explorers, seeing these buildings from a distance, into believing they had found the fabled Seven Cities of Gold. One sees what one wants to see. Si?

If one is going to plaster an adobe building there is a process. Plaster over adobe has to have chicken wire to have the final coat adhere. Here the original adobe shows, then the plaster, then the chicken wire waiting for the next adobe finish.

ARCHITECTURE AND INTERIORS
PATIOS

At the high altitude of northern New Mexico summers are short, so people move outside to enjoy the season as many hours as possible. Patios and *portales* are important parts of homes and commercial establishments.

All types of handiwork are brought together on the patio: brick floor, beams, posts, corbels, handmade furniture and artwork.

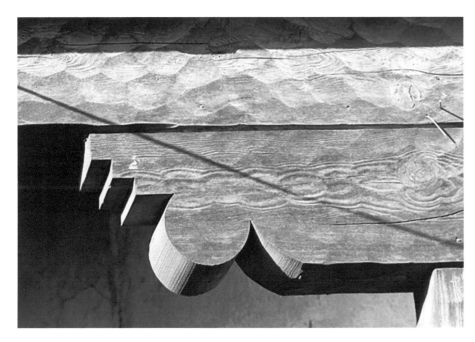

Frank Romero, from the Taos Pueblo, executed the adze work on the beam and corbel at the Kit Carson House in Taos.

Willa Cather wrote *Death Comes to the Archbishop*. Her protagonist, Archbishop Lamy, built this house at 300 Galisteo Road in Santa Fe, for his organist, Fransisca Hinojos, prior to 1885. A plaque reads, "The Santa Fe Foundation finds this building worthy of preservation."

The beam on the patio was moved from the Nambe mission around 1900. The inscription says, "This Church was built by Governor General Bustamante for His Majesty in the year 1725." The corbel, made from two large squared logs, is double the usual size.

Delicate painting of vines flanks the door to the entry patio with coyote fence beyond.

PUERTAS / DOORS

The first doors at the pueblos were holes in the ceilings reached by ladders that could be drawn in when enemies were approaching. When times became more peaceful doors in front walls came into use. Doors can be humble, ecclesiastical, weighty, filigreed, ponderous, recycled, stock grade or artist made.

Doors are sometimes welcoming and sometimes a challenge, as this one is. You have to both step up and watch your head.

Some of the doors shown in this section of the book were made in this workroom. The door in the back and the detail on the column in the foreground are some examples of the output of the artist who works here. The tools are quiet now, but there is a whir and grind when work is underway.

The lintel is narrower than the door, as though it could not be doing its job.

This screen door is not one you find at your local hardware store. You smile even before you go inside. Jose Dolores Lopez was a carpenter from Cordova, New Mexico, who lived from 1868 to 1937. An example of his work is shown in the Smithsonian American Art Museum.

Iron strap work for the extended hinges includes a sophisticated locking mechanism.

"Spindles" seems like an inadequate word to describe the components of this door, but what else is there to say?

The multiple chevrons in a series of doors emphasizes the theme of repetition. Repeated components of buildings give evidence of the craftsmanship expressed wherever possible.

Corn is an important crop in the world influenced by Indians, so it makes sense as a design for a pantry door.

The cross in this door at the Millicent Rogers Museum in Taos is probably a reference to a Navajo design.

The fine craftsmanship of these shed doors are only seen by the owners of the enclosed back yard.

The zipper door is aptly named and illustrates skill in its design and execution.

So much to see in this home. The open door shows unusual wood detail and glass and iron. This ornate mirror hangs on a wall illustrating plaster trowel work. The reflection showing the two steps down means the room is draft-free.

Coming home is always interesting and guests feel special before they even enter.

Note the carved shell design at the bottom of this door.

The door ajar is in the Randall Davey home which is now a museum. Davey was a member of the Santa Fe Artists Society, and was recognized as a print maker, painter, and sculptor. The property has been donated for a wildlife sanctuary.

Shhhh. Could this be a door to a nursery?

The Moroccan star is the design of this door at the Dar Al-Islam Foundation School.

48

An open doorway is a frame. The doorway itself is subject and object. It frames something. The eye must focus up close, but also through and beyond. Past the room is a private chapel open to the sky.

Originally windows were small and few. Homes snuggled shoulder to shoulder around the plaza. First there were small slits in the wall to let in a little light and to serve as ports for shotguns when under attack. The next iteration for the windows, before wagon trains brought glass, was to cover the opening with hides, followed by mica—a naturally occurring translucent mineral. Despite the vibrant light outside, life inside must have been claustrophobic.

Windows illustrate how deep the walls are, which usually distinguishes one of the differences between old and new construction. The walls were great insulation, keeping people warm inside in the winter and cool when it was hot outside. But with today's central heat and the cost of labor making the thick walls being very expensive, walls now are thinner. Deep windowsills show either age or wealth.

Betty Stewart is the contractor and owner of the home with this window built with 10 inches plus air space for insulation and another 10 inches of adobe brick. The brick usually is 10" x 7" x 4" and can be laid with the 10" dimension going from inside to outside instead of laterally. Adobe allows remodeling such that windows can be added later.

In *Roots in Adobe,* Dorothy L. Pillsbury writes, "When I bought the Little Adobe House years ago, it had a long narrow kitchen with a blank wall on one side...Why not fill that wall space with all the little square paned windows it would take?...I might have six feet of window glass over a yard high fence facing the southern sky...they worked hacking out the thick adobe brick carefully. By noon...I owned a flood of golden sunlight."

Adobe walls range from one to two bricks thick. The bricks are 17" x 10" x 7". The bricks can be laid with the longest side lying in and out. With the glazing placed toward the outside, the sills are more than deep enough for the ubiquitous geranium plants.

This window surround has the signature of Betty Stewart, a contractor who was building in the 1980s. The precise edging and the polished finish that were her signature show up in the top left of the window frame. Her work was sought after by people who could afford the thick walls and meticulous detailing. Every Santo (figure of a saint) like the one seen outside the window confers a specific blessing. Something in the clothing or an attached item will made the identity clear, but we can't see this one clearly enough to guess.

The board on the left of the window helps illustrate the depth of the wall.

The *euphorbia trigona* plants make striking drapery panels. Slow growing, they may be as old as the room.

The humble cock-eyed window contrasts with the wealth evident in the use of appropriate building materials, nice furnishings and idiomatic accessories.

Grillwork protecting windows like this classic federal style window frustrated author Frank Waters, who had his protagonist in *The Man Who Killed the Deer* mutter, "The windows were stripped with iron, the best protection against washing."

CHIMENEA / FIREPLACES

What is a home without a hearth? In New Mexico, as in many places, the climate is such that it can be chilly any time of the year. How welcoming is a fire in the fireplace! Fireplaces are attractive when cool and quiet but native piñon pine, when lit, wafts a wonderful smell and crackles brightly. The smoke lifting to the stars lets anyone outside imagine that all is well inside.

Fire was first in a ring of stones, then in an outdoor oven and here it is moving closer to inside. Next the fire was built in the corner of the room under an opening in the ceiling. A pipe might have been in place to take the smoke through the ceiling and roof. Inside it usually remains in the corner, with wood stacked as it would have been at an outdoor campfire.

Anita Rodriguez, whose profession it was to build with adobe and to lay mud floors, instructed me to write, "A fireplace is the perfect androgynous symbol: The fireplace is the female form—the vessel.

In the fireplace are combined earth, water, air and fire. As the fire burns the smoke goes up as spirit. The chimney is the phallus."

She added that it is important that the fireplace symbolize the woman. It is the center of the home and where all the food was cooked all winter.

Fireplaces are usually located in the corner to radiate heat into the room. The shape is called a "beehive." Also shown on the back wall is the look of paint applied directly on the adobes. The tile floor is typical.

This fireplace is a classic. It is such a simple object to give so much comfort. No andirons are used. The logs stand on end leaning against one another. They give the effect of the outdoor campfire, moved to the corner of the room. There is an infinite amount of variation, but it is this form, the beehive, from which all others deviate.

"Necessary rooms" like kitchens and bathrooms are made better with fireplaces.

Every inch is used to full advantage in this bathroom.

This fireplace has more detailing but, as most fireplaces in the area, has the same oval opening. The Arts and Crafts furniture is arranged for a sociable evening.

Boris Gilbertson has this collection of his art and that of his friends displayed by his fireplace.

There can be more to a fireplace than a pretty face and flue. The shepherd's bed is meant to capture the heat retained in the fireplace surround and to use it for toasty sleeping. Originally this would have created a sleeping area in a one-room home.

From the shepherd's bed comes the idea for a platform, a bit too short for sleeping, but making a good display shelf. The supports at the front edges make room for a bench below.

The fireplace designed by Frank Lloyd Wright reinforces the idea of an Indian vessel in his Pottery House.

This grand room has the traditional components of mud floors, adobe walls, corner *fagon* and ceiling supported by *vigas*. Randall Davey imposed his European ideas with the antique furnishings, fireplace mantelpiece and shuttered windows, which bring a sophistication unusual in adobe homes. The formal fireplace front is juxtaposed onto an adobe chimney.

Forrest Fenn's fireplace in the guest apartment of the gallery he owned in Santa Fe strongly states presence. After closing his gallery Fenn shook things up by publishing a number of clues that he said would lead to a treasure worth about three million dollars. He had hidden a bronze chest weighted down with about 40 pounds of gold, precious gems and pre-Columbian figures that would make someone wealthy. In 2016 a man died searching. He called a friend on a cold January evening, with snow about to fall, saying he thought he was close. No word after that and his body was later found. Before the treasure was found in Wyoming a second man died. Soon after the treasure was found in 2020, Fenn died.

Here are the clues:

As I have gone alone in there
And with my treasures bold
I can keep secrets where,
And a hint of riches new and old
Begin it where warm waters halt
And take it in the canyon down.
Not far, but too far to walk
Put in below the home of Brown.
From there it's no place for the meek,
The end is drawing ever nigh;
There'll be no paddle up your creek.
Just heavy loads and water high.
If you've been wise and found the blaze,
Look quickly down, your quest to cease.
But tarry scant with marvel gaze,
Just take the chest and go in peace.
So why is it that I must go
And leave my trove for all to seek?
The answers I already know
I've done it tired, and now I'm weak
So hear me and listen good,
Your efforts will be worth the cold.
If you are brave and in the wood
I give you title to the gold.

PISO / FLOORS

The humble mud floors have elitist qualities. Because the labor is intensive, what was once a poor person's necessity is now a rich man's trophy. Dirt floors offer challenges as described by Frank Waters in *Woman of Otowi Crossing*: "...he contritely hopped about mopping the floor, having been too long a bachelor housekeeper to let the water, even now, destroy his hard-won glaze on the earthen floor." In *People of the Valley* he writes, "Patches of dirt floor were hard and smooth and black with the blood mixed with adobe."

Anita Rodriguez is an expert in finishing mud surfaces and had the following instructions for coating the floor in a course I once took from her:

"There are many recipes for mixing the mud with something like the blood which may have been from a goat. *Tierra Blanca* (white earth) found in nature, mixed with a teaspoon of cream per gallon of flour paste is one recipe. Some sources mention milk or Elmer's glue.

"Mix 1 gallon of carpenters glue with 6, 7, or 8 parts of water poured from the same container. Stir constantly or the glue will settle. Add the same proportion of mud with this glue mixture to reach a dry consistency that still holds together."

or:

"With a mixture of 3/4 linseed oil to 1/4-part kerosene, heat until you can barely touch with your finger. Add five coats of this to the floor. After each application has dried increase the portion of kerosene until it is almost all kerosene, the inverse of the above proportions.

"The first coat dries fast. The second coat can go on when the first coat is no longer tacky or shiny. The next morning add the 3rd coat, the 4th coat that afternoon and the 5th coat the next day. Then allow a day or two for all to set."

This kitchen is owned by Ms. Rodriguez who made it her mission to preserve the art of laying a mud floor. She is an *enjaradora*, a mud specialist, who builds fireplaces and horno ovens, as well as replastering adobe buildings. She held workshops to forestall the economic and social consequences should this craft be lost.

Dark brown mud floor is a sharp contrast to the white plastered walls.

The home again of Boris Gilbertson with a focus on the mud floor. He used to let in the public and mud passed the test of heavy traffic. The adobe exterior is a sharp contrast to the formality of the room. It lacks a finish coat so one sees the rough nature of adobe mixed with straw.

Formerly the home belonged to Donaciano Vigil, who served as acting governor of the Territory of New Mexico. Built circa 1832 the house has been carefully restored by subsequent owners who donated the house, orchards and farmlands to be preserved and open to the public.

The owner laid herringbone brick floor and used sand as fill. She also made the iron sculpture and handholds at the stairs.

This is the classic *viga* and *latilla* ceiling.

TECHO / CEILINGS

Ceilings in adobe buildings are most typically made of *vigas* or logs spaced a few feet apart. *Vigas* are large beams that rest on the tops of walls and protrude beyond. Between the *vigas* are *latillas*—usually willow twigs or cedar slats or even milled lumber. This way of building was born of necessity and remains because of its esthetic richness.

In the old days a man would take his mule into the hills, cut down two trees the same size and fasten them to either side of his mule's saddle. In order to secure the logs, he would burn a groove in the logs with a branding iron and attach them with a rope so they wouldn't slide off. As the man and beast walked home the ends dragging, the logs wore down to a point. One may occasionally see pointed *vigas* protruding from a wall. Most trees are substantial enough to serve as *vigas* for about 15 feet, which dictates that the width of the room is less than 15 feet because the *vigas* protrude beyond the walls.

The *latillas* fill the space between *vigas* and are much smaller. They might be small branches or willow twigs. Stripped of bark they are let to dry. Very many *latillas* were needed in this room so note again how much hand labor is involved.

Square beams with sturdy corbels that hold up flat lumber in place of more traditional *latillas* are seen in the Southwest Spanish Craftsmen House and Showroom in Santa Fe.

The ladies in this former bordello expressed their femininity with lace on the *vigas*. This is now a private home.

The ceiling in this room of the Southwest Craftsmen House has borne a heavy load of sod for a very long time. Right behind the chain of the chandelier is a crack in one of the *vigas*. With age comes osteoporosis. The geraniums form a better window covering than the draperies now retired to each side. Near the vestibule the wall on either side has a ribbing in the plaster, implying pillars.

Scaling up is this ceiling which has beams that are pine logs—much larger than usual. The *latillas* are themselves small trees. This home is as much a museum as a house. The significant collection of pots and shields are noteworthy. On the top shelf at the far left is a Navajo man's hat.

This strong ceiling is in the bed/sitting room of the Fenn Gallery guest wing in Santa Fe.

The ceiling is very strong even in the bathroom.

While it is hard to tell new from old in Adobe Country, this tin ceiling, if old, would be evidence of commerce from the east and of enough wealth to make the purchase.

Hasan Fathy, Egyptian architect of international repute, designed a mosque in New Mexico for the Dar Al Islam Foundation in 1985. Dar Al Islam means Place of Peace. Native craftsmen with years of experience working with adobe got to build their first domed ceiling. The foundation in Abiquiu, New Mexico also has a school and library, each with a dome leading up to a small window.

A small window in time allowed for this photograph showing the attractive pattern of brick before the plaster covered up the brick. The bricks are smaller than usual and appear to get even smaller as the courses rise to the top.

TECHO / ROOFS

John Nichols, in *If Mountains Die, a New Mexico Memoir*, writes, "To make double sure all the indigenous flat-topped roofs like mine leak profusely drainage is provided by one or two tin *canales* (downspouts) protruding from the fire wall. These are usually blocked by ice, or by leaves and twigs blown off the surrounding cottonwoods, hence their real historical function is to impede, rather than facilitate, drainage."

Flat roofs rimmed by firewalls are a maintenance challenge as they collect and hold the snow that softens the adobe. For greater insulation the roof may be covered with sod. Here at the Taos Pueblo the sod is producing vegetation.

Flat roofs are rimmed by firewalls.

The consistency of detailing of this Territorial home: stucco to carry on the adobe look, the tin roof, the bit of gingerbread on the porch and door trim make this house a Victorian jewel.

The peaked attic is guarded by the gargoyle at the end of the center beam. He is ready to go to work with the ring in his mouth that can hoist a rope. Metal roofs such as this are either the natural tin color or orange.

The generous attic space creates an ecological HVAC system. In the summer the trapped air under the roof is terribly hot. By opening the dormers hot air escapes and by opening a door to the rooms below the hot air is sucked up and out, leaving the rooms below almost too cold. In the winter the dormers are closed; the trapped air still heats up and a fan pulls the warm air down into the house. The wood carving of the porch rail supports shows local artisan talent. Don Quixote may be saying that it is not an impossible dream about living well adobe style.

At Christmas luminarias mark roof lines and walkways. This many are surely electrified, but the authentic way to decorate is to fill brown paper bags with some sand, which safely hold the lighted candle that lasts only the one night and is replaced the next day. Especially lovely to see are these simple lanterns when there is snow on the ground. Carolers are hoping that when they see the lighted luminarias there will be an offer of hot chocolate and cinnamon flavored cookies, *biscochitos*.

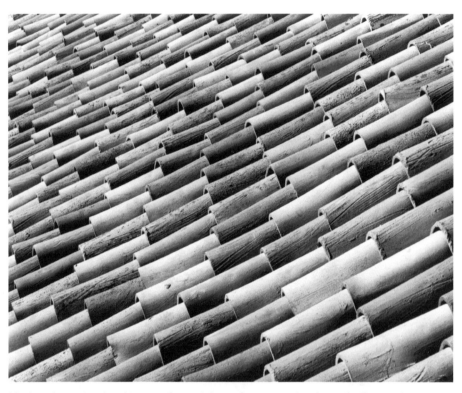

Mud roofs are too hard to care for and tin roofs may not be the esthetic people want on newer homes. Tile fits the southwest genre.

COLOR

Adobe, coming from the earth, is brown, but for detailing there is a surprising number of colors that come from places where people know to look. In *Adobe Homes and Interiors of Taos, Santa Fe and the Southwest*, by Sandra and Laurel Seth, a home is shown with mauve on the walls. The plaster's dirt came from the Sandia Mountains.

Window and door frames are often painted blue, which in many Catholic countries represents the Virgin Mary. The color is thought to act as a blessing and to offer protection to those within. The blue seen so often in northern New Mexico is even called Taos Blue.
Skies in New Mexico are usually intensely blue, giving a strong background to the dun-colored ground and homes.

White paint applied directly on the brown adobe around doors and windows with their blue frames makes a strong statement. Sometimes the upper quadrant will be painted with a flourish of curves. Formerly, the white paint was finely ground gypsum to which wheat flour had been added to form a paint which was applied to the wall with a piece of sheepskin.

Mabel Dodge Luhan, in *Winter in Taos*, describes her large dining room as "all in light (west facing windows) with walls whitewashed in pale grey earth..." Mabel herself was "...a director of human display, who created the settings, gathered brilliant actors and activated the drama." This description is by John Collier, Jr. in the foreword to *Edge of Taos Desert*, by Mabel Dodge Luhan.

THE MUD HUT NUTS

Helen Doyle pulled "mud hut nuts" from "A Calendar of Santa Fe", a short story by Winfield Townley Scott, from a compilation of short stories titled *The Spell of New Mexico,* edited by Tony Hillerman.

Artists who came to Taos and Santa Fe found a way of living day to day that was a departure from what they had known back east and what they had expected to find. Their admiration for adobe architecture led to advocating for building codes that require buildings in Santa Fe to have the adobe look.

John Gaw Meem, architect, was commissioned to design many important pueblo and colonial style buildings that followed the 1957 codes. His work is evident on the plaza of Santa Fe and, notably, on the University of New Mexico campus in Albuquerque. There eighteen substantial commissions and several remodels are his. The Zimmerman Library is said to be his most beautiful. Also, in Taos, the Presbyterian Church is the simple and lovely example of his work. Bainbridge Bunting's book, *John Gaw Meem, Southwestern Architect,* has many more illustrations of his work.

Dorothy L. Pillsbury, in *Adobe Doorways*, remarks that most Pueblo Indians and Spanish Americans (1952) live in adobe houses. "Among the Anglos, a veritable cult of adobe dwellers has sprung up."

GLOSSARY

aborigine: first inhabitants contacted by invading or colonizing people

bond beam: horizontal structural element giving strength to a wall typically providing a consistent anchorage for a roof structure

conquistadors: soldiers from Spain and Portugal out to conquer new territory

coyote fence: fence made of uniform sticks tall enough to keep out predators

faralitos: brown paper bags with sand and candles within, lit especially at Christmas time, and used to line rooflines and paths

horno: outdoor oven with a 1' x 1' opening and a vent hole, usually fired with cedar wood

latillas: peeled branches or wood used between beams

luminaries: interchangeable with faralitos or meaning small bon fires

mica: sheet silicate found in the Manzano Mountains in New Mexico

micaceous clay: dirt with bits of mica throughout that adds sparkle

nicho: An imbedded box in a wall that might hold a statuette of a saint or a candle.

plaza: open space surrounded by buildings, public marketplace

poco tiempo: a little time or not long

pueblo: village, an American Indian village of Arizona or New Mexico of flat roofed adobe houses joined in groups or, when capitalized, an American Indian people of the Southwest United States

santero: artisan who makes religious art, an art form of Colorado and New Mexico

tierra blanca: white earth

treading: trampling with feet, in this case repeatedly to make dirt dense

vigas: wooden beams that carry the weight of the roof

zaguan gate: a gate that has a door for people within a gate large enough for wagons

SUGGESTED READINGS

Adams, Robert. *The Architecture and Art of Early Hispanic Colorado*, Colorado Associated University Press in Cooperation with The Historical Society of Colorado, 1974.

Anaya, Rudolfo A. *Bless Me, Ultima*, A Publication of Tonatiuh International, 1972.

———. *The Silence of the Llano*, Tonatiuh-Quinto Sol International, 1982.

———. *Tortuga*, University of New Mexico Press, 1979.

The Anschutz Collection. *Painters and the American West*, Yale University Press, 2001.

Baca, Elmo and Suzanne Deats. *Santa Fe Design*, Publications International, 1990.

Buchanan, Ken. *This House Is Made of Mud,* Northland Publishing, 1991.

Bunting, Bainbridge. *John Gaw Meem*, University of New Mexico Press, 1983.

———. *Taos Adobe,* Museum of New Mexico Press, 1964.

Burba, Nora and Paula Panich. *The Desert Southwest*, Bantam Books, 1987.

Cather, Willa. *Death Comes to the Archbishop*, Modern Library, 1931.

———. *My Antonia*, Houghton Mifflin Company, 1949.

———. *O Pioneers*, Houghton Mifflin Company, 1913.

Childs, Craig. *House of Rain*, Little, Brown and Company, 2006.

Church, Peggy Pond. *The House at Otowi Bridge,* University of New Mexico Press, 1959.

Clark, H. Jackson. *The Owl in Monument Canyon*, University of Utah Press, 1993.

Colby, Catherine. *Kate Chapman, Adobe Builder in 1930s Santa Fe*, Sunstone Press, 2012.

Davis, Christopher. *North American Indian*, The Hamlyn Publishing Group, Limited, 1969.

De Borhegyti, Stephen F. *El Santuario de Chimayo*, The Spanish Colonial Arts Society, Inc., 1956.

Dennis, Landt. *Behind Adobe Walls*, Chronicle Books, 1997.

The Denver Art Museum. *Picturesque Images from Taos and Santa Fe*, 1974.

The Denver Public Library. *Pueblo Treasure*, 2005.

De la Baca, Vincent C., editor. *La Gente*, Colorado Historical Society, 1998.

D'Emilio, Sandra and Susan Campbell. *Images of Ranchos de Taos Church*, The Museum of New Mexico Press, 1987.

Drain, Thomas A. *A Sense of Mission*, Chronicle Books 1994.

Eastlake, William. *The Bronc People*, University of New Mexico Press, 1957.

Emmerling, Mary. *Art of the Cross*, Gibbs Smith Publisher, 2006.

Erdoes, Richard and Alfonso Ortiz. *American Indian Myths and Legends*, Pantheon Books, 1984.

Exhibition at Centre Georges Pompidou. *Down to Earth*, Facts on File, 1981.

Gray, Virginia and Alan Macrae. *Mud Space and Spirit*, Capra Press, 1976.

Hammett, Jerilou, Kingsley Hammett, Peter Schollz. *The Essence of Santa Fe*, Ancient City Press, 2006.

Hillerman, Tony, ed. *The Best of the West*, Harper Perennial, 1991.

———, ed. *The Spell of New Mexico*, University of New Mexico Press, 1976.

Hooker, Van Dorn. *Centuries of Hands, An Architectural History of St. Francis of Assisi Church in Taos, New Mexico*, Sunstone Press, 1996.

74

Hunt, Edward Proctor. *The Origin Myth of Acoma Pueblo*, Hunt Edward Proctor, Penguin Books, Unpublished proof.

Iowa, Jerome. *Ageless Adobe, History and Preservation in American Southwestern Architecture*, Sunstone Press, 1998.

Keegan, Marcia. *Pueblo People*, Clear Light Publishers, 1998.

Kloss, Phillips. *The Taos Crescent*, Sunstone Press, 1991.

Kutsche, Paul and John R. VanNess. *Canones*, University of New Mexico Press, 1981.

La Farge, Oliver. *Behind the Mountains,* Sunstone Press (New Edition, 2008).

———. *The Door in the Wall*, Houghton Mifflin Company, 1965.

———. *The Man with the Calabash Pipe,* Sunstone Press (New Edition, 2011).

Locks, Raymond Friday. *Sweet Salt,* Roundtable Publishing, 1990.

Luhan, Mabel Dodge. *Edge of Taos Desert*, University of New Mexico Press, 1937.

———. *Winter in Taos*, Sunstone Press (New Edition, 2007).

Lummis, Charles F. *The Land of Poco Tiempo*, Lummis Charles F., University of New Mexico Press, 1952.

Luther, T.N. *Collecting Taos Authors*, New Mexico Book League, 1993.

Marriott, Alice. *Maria: The Potter of San Ildefonso*, University of Oklahoma Press, 1948.

Mather, Christine and Sharon Woods. *Santa Fe Houses*, Clarkson Potter, 2002.

———. *Santa Fe Style,* Rizzoli, 1986.

McHenry, Paul Graham Jr. *Adobe and Rammed Earth Buildings,* John Wiley and Sons, 1984.

Morrill, Claire. *A Taos Mosaic*, University of New Mexico Press, 1973

Nader, Khalili. *Racing Alone*, Harper and Row, Publishers, 1983.

Neihardt, John G. *Black Elk Speaks* University of Nebraska Press, 1932.

Pillsbury, Dorothy. *No High Adobe,* University of New Mexico Press, 1950.

———. *Roots in Adobe*, University of New Mexico Press, 1959.

Rakocy, Bill. *Taos Artists Founders*, Bravo Press, 2002.

Reeves, Agnes. *The Small Adobe House,* Gibbs Smith Publisher, 2001.

Sanchez, Alex and Laura Sanchez. *Adobe Houses for Today*, Sunstone Press, 2003.

Scott, Sascha T. *A Strange Mixture,* University of Oklahoma Press, 2015.

Seth, Sandra and Laurel. *Adobe,* Taylor Trade Publishing, 1988.

Spears, Beverley. *American Adobes*, University of New Mexico Press,1986.

Stedman, Myrtle. *Adobe Architecture*, Sunstone Press, 1987.

———. *Adobe Remodeling and Fireplaces*, Sunstone Press, 1986.

——— and Wilfred Stedman. *Artists in Adobe*, Sunstone Press, 1993.

Stuart, David. *Anasazi America*, University of New Mexico Press, 2000.

Time Life Books, Editors. *The Spanish West,* Time-Life Books, 1976.

Transcriptions of Oral Literature, The Zunis, The New American Library, 1972.

Ulibarri, Sabine. *Tierra Amarilla*, University of New Mexico Press, 1971.

Warren, Nancy Hunter. *New Mexico Style*, Museum of New Mexico Press, 1986.

Waters, Frank. *Leon Gaspard,* Fenn Galleries, Ltd. / Northland Press, 1981.

Witynski, Karen, and Joe P. Carr. *Adobe Details*, Gibbs Smith, 2002.

Wood, Nancy. *Many Winters,* Doubleday and Company, Inc., 1974.

PEOPLE WHO HELPED

Jean Adams

Betty Angelos Sabo

Jeff Arens

Tony Armijo

Ron Baird

Betty Berry

Bill Branch

Bainbridge Bunting

Robert Burdsal

Jim Cartwright

Peter and Alicia Cavritzen

Max Chavez

Adrian Christian

Dorian Christian

Betty Colbert

John Conran

Tom Carr

Bruce Cousins

Jo Diggs

Jack Edwards

Mr. and Mrs. William M. Field

Mike Freeman

Pat French

Marilyn Foss

Kelly M. Graves

Virginia Grey

Pete and Francie Handler

Bob Harcourt

Pat Harrison

Genevieve Janssen

Kyle Holman

Tim Jelinek

Jean Jennings

Betty Jones

Elle Jones

Starr Jones

Arthur and Kate Kent

Roy and Beverly Krosky

Jody La Fon

Weebelos La Salle

Patricia R. Loree

Jim and Fran Mack

Jean Martin

Joanna Martin

Christine Mather

John R. McGill

Caryl McHarney

Greg and Mike Martin

Patrick and Joanne Martin

Jim and Patricia Martindale

Dan Martinez and Dennis Garcia

Ida Martinez

John and Fran Marvel

Margaret Maupin

Joan McLaughlin

Robert and Betty Meader

James Messimer

Roger and Ellen Miller

Barbara Money

Nick Morrow

Barnique Longley Orr

Forrest Moses

Sam Murray

Nat Owings

Nino Padilla

Jan Parker

Nance Parker

Brett Patrick

Karen Paul

Jerry and Katie Peters

Walter Pickett and Glenn Panebeauf

Judy Pollock

Neil Posse

Peggy Reed

Matthew Restall

Pablo and Hernando Rivera

Joel Russman

Zoey Ryerson

Ralph and Katherine Schomp

Dana and Mary Schriber

Sandra Seth

David P. Stanley

Duke Sundt

Pamela Telleen

Tom Thomason

Vince Vigil

Mike Vogel

Ted Waddell

Sally Wagner

Billie Walters

Lucy Ann Warner

Paulette, Kara, Kristen Lindsey, Bryn and Lauren Weaver

David and Lydia Miller

Theodor H. Waddell

Charlotte White and Boris Gilbertson

Quentin Wilson

Cathy Wright

Donald Wright

ABOUT THE AUTHOR

Marcia Johnson was born and grew up in Alamosa, Colorado in the San Luis Valley in view of the Sangre de Cristo Mountains, which is the same range that backstops northern New Mexico towns and villages. The San Luis Valley in Colorado is the Taos Valley once one crosses the state line south into New Mexico.

Marcia has returned over and over to the areas covered in the book, drawn to their qualities much as the Taos artists were. Her library is well stocked with fiction and non-fiction set in northern New Mexico.

She has a BA in philosophy *cum laude* from The Colorado College.

ABOUT THE PHOTOGRAPHER

Michael Gamer trained in photography as an apprentice to well-known Myron Wood and in commercial studios in Denver. Michael also studied with Ernst Haas, Jay Maisel and Eliot Porter. After liberal arts studies at The Colorado College he continued his education in the fine arts department of the University of Colorado.

Michael spent a year as a staff photographer for the Navajo Tribal Museum in Window Rock, Arizona. His work has been exhibited in art shows throughout the West and is held in private collections in US West, now CenturyLink, and the United Banks of Colorado, now Wells Fargo. Additional trips to Mexico, Guatemala and Peru stimulated his vision as a photographer as well as a student of anthropology. Some of his photographs are archived in the Western History and Genealogy Department of the Denver Public Library.

CPSIA information can be obtained
at www.ICGtesting.com
Printed in the USA
BVHW052252261021
619790BV00001B/2

9 7 8 1 6 3 2 9 3 3 5 3 9